ADDRESS AT THE MEETING
CELEBRATING THE 25TH ANNIVERSARY OF
HONG KONG'S RETURN TO THE MOTHERLAND AND
THE INAUGURAL CEREMONY OF
THE SIXTH-TERM GOVERNMENT OF
THE HONG KONG SPECIAL ADMINISTRATIVE REGION

1 July, 2022

XI JINPING

U0132273

ADDRESS AT THE MEETING CELEBRATING THE 25TH ANNIVERSARY OF
HONG KONG'S RETURN TO THE MOTHERLAND AND
THE INAUGURAL CEREMONY OF THE SIXTH-TERM GOVERNMENT OF
THE HONG KONG SPECIAL ADMINISTRATIVE REGION

Author   Xi Jinping

Published by Joint Publishing (H.K.) Co., Ltd.
20/F., North Point Industrial Building, 499 King's Road, North Point, Hong Kong

Distributed by SUP Publishing Logistics (H.K.) Ltd.
16/F., 220-248 Texaco Road, Tsuen Wan, N.T., Hong Kong

Printed by Elegance Printing & Book Binding Co., Ltd.
Block A, 4/F., 6 Wing Yip Street, Kwun Tong, KLN, Hong Kong

First Published in July 2022
ISBN 978-962-04-5062-4

Address at the Meeting Celebrating the 25th Anniversary of Hong Kong's Return to the Motherland and the Inaugural Ceremony of the Sixth-Term Government of the Hong Hong Special Administrative Region

1 July, 2022

Xi Jinping

Fellow Compatriots,

Dear Friends,

Today we are gathered here to celebrate this grand occasion marking the 25th anniversary of Hong Kong's return to the motherland, and to hold the inaugural ceremony of the sixth-term government of the Hong Kong Special Administrative Region (HKSAR).

First of all, I would like to extend sincere greetings to all the people of Hong Kong. I also extend warm congratulations to the newly inaugurated sixth-term HKSAR Chief Executive Mr. John Lee, principal officials of the sixth-term HKSAR government, and members of the Executive Council. And let me express heartfelt appreciation to all our compatriots both at home and abroad, and international friends for their support for the cause of "one country, two systems" and for Hong Kong's prosperity and stability.

In the over 5,000-year history of Chinese civilization, our ancestors working hard on the land south of the Five Ridges is an important chapter. In the history of modern China after the Opium War in 1840, the humiliation of ceding Hong Kong is a page of pain, and also included is the Chinese people's fight for the survival of our country. The past century has witnessed how the Communist Party of China has united and led the Chinese people in its magnificent endeavors for a better future, to which fellow Chinese in Hong Kong have made unique and significant contributions. Throughout history, people in Hong Kong have always maintained a close bond with the motherland in weal and woe.

Hong Kong's return to the motherland marked the beginning of a new era for the region. Over the past 25 years, with the full support of the country and the joint efforts of the HKSAR government and people from all walks of life in Hong Kong, the success of "one country, two systems" has won recognition throughout the world.

Since its return, Hong Kong, amid China's monumental

reform and opening-up efforts, has been breaking new ground, functioning as an important bridge between the Chinese mainland and the rest of the world. As a result, it has made irreplaceable contributions to our country's economic miracle marked by long-term, stable, and rapid growth. Proactively integrating itself into the country's overall development and carving out its role in national strategies, Hong Kong has maintained its strengths in its high degree of openness and in aligning with international rules. In doing so, the region has been playing an important role in raising China's opening up to a higher level with wider coverage and scope. With continuously expanding areas and enabling mechanisms for Hong Kong's cooperation and exchanges with the mainland, people in Hong Kong now have better opportunities to start their own businesses and make achievements.

Since its return, Hong Kong has overcome various hardships and challenges and advanced steadily forward. Be it the global financial crisis, the COVID-19 epidemic, or social unrest, none of them have stopped Hong Kong from marching forward. Over the past 25 years, Hong Kong's economy has been thriving, its status as an international financial, shipping, and trading center has been maintained, and its innovative science and technology industries have been booming. Hong Kong has remained one of the most liberal and open economies in the world, it has also maintained a world-class business environment, its previous laws including the practice of the common law have been maintained and developed, various social programs have made all-around progress, and overall social stability has been ensured. As a cosmopolis, Hong Kong's vitality has impressed the world.

Since its return, Hong Kong has ensured its people enjoy their status as masters of the region. With the policy of the people of Hong Kong administering Hong Kong and a high degree of autonomy in practice, the region has truly entered an era of democracy. Over the past 25 years, Hong Kong's constitutional order based on the Constitution of the People's Republic of China and the Basic Law of the HKSAR has been maintained in a steady and sound manner. The central government's overall jurisdiction over Hong Kong has been well implemented, and a high degree of autonomy in the region has been exercised as it should. The Hong Kong National Security Law was adopted, which has established the legal system to safeguard national security in the region. The electoral system of Hong Kong has been modified and improved, thereby materializing the principle that Hong Kong should be administered by patriots. The democratic system of the special administrative region (SAR) conforms to both the "one country, two systems" principle and the region's constitutional status. It is in the interest of Hong Kong residents' democratic rights and the region's prosperity and stability, securing a bright future for the region.

Fellow Compatriots,

Dear Friends,

"One country, two systems" is an unprecedented innovation. Its fundamental purpose is to safeguard China's sovereignty, security, and development interests and to maintain long-term prosperity and stability in Hong Kong and Macao. All that the central government has done are for the benefits of Hong Kong and Macao, for the well-being of all residents of the two regions, and for the future of the whole country. At the meeting celebrating the 20th anniversary of Hong Kong's return to the motherland in 2017, I stated that the central government, in implementing the principle of "one country, two systems," will maintain two key points: first, the central government will remain resolute in implementing the principle, and will not change or vacillate in this stand; and second, the principle will be implemented as what it is originally intended precisely. Today, I would like to stress again that "one country, two systems" has been tested repeatedly in practice. It serves the fundamental interests of not only Hong Kong and Macao, but also the whole country and the nation. It has gained wide support from the 1.4 billion-plus Chinese people including the residents of Hong Kong and Macao. It is also widely accepted by the international community. There is no reason for us to change such a good policy, and we must adhere to it in the long run.

Fellow Compatriots,

Dear Friends,

A review of the past can light the way forward. The practice of "one country, two systems" in Hong Kong has left us both valuable experience and profound inspiration. What has been done over the past 25 years tells us that only if we have a profound and accurate understanding of the laws guiding the practice of "one country, two systems," can we make sure our cause advances in the right direction in a sound and sustained manner.

First, we must fully and faithfully implement the principle of "one country, two systems." This principle embodies a complete system. Its top priority is to safeguard national sovereignty, security, and development interests. With this as a prerequisite, Hong Kong and Macao can keep the previous capitalist systems unchanged for a long time and enjoy a high degree of autonomy. Since the socialist system is the fundamental system of the People's Republic of China and leadership by the Communist Party of China is the defining feature of socialism with Chinese characteristics, all residents in the special administrative regions should willingly respect and uphold the country's fundamental system. The thorough and precise implementation of the "one country, two systems" principle will open up broader prospects for the development of Hong Kong and Macao. The more firmly the "one country" principle is upheld, the greater strength the "two systems" will be unleashed for the development of the SARs.

Second, we must uphold the central government's overall jurisdiction while securing the SARs' high degree of autonomy. Since Hong Kong's return to the motherland, it has been re-integrated into China's governance system, and a constitutional order was established with the "one country, two systems" principle as its fundamental guideline. The central government's overall jurisdiction over the SARs underpins their high degree of autonomy, and such autonomy bestowed by the law is fully respected and resolutely safeguarded by the central government. Only when the enforcement of the central government's overall jurisdiction dovetails with the fulfillment of a high degree of autonomy in the SARs, can the SARs be well governed. The SARs uphold the executive-led system. The executive, legislative, and judicial branches perform their duties in accordance with the basic laws and other relevant laws. The executive and legislative branches check and balance and cooperate with each other while the judiciary exercises its power independently.

Third, we must ensure that Hong Kong is administered by patriots. It is a universal political rule that a government must be in the hands of patriots. There is no country or region in the world where its people will allow an unpatriotic or even treasonous force or figure to take power. The government of the HKSAR must be safely kept in the hands of those who love the country. This is an essential requirement for Hong Kong's long-term prosperity and stability and must not be compromised under any circumstances. To put the governing power in the right hands is to safeguard Hong Kong's prosperity and stability as well as the immediate interests of more than 7 million people in the region.

Fourth, we must maintain Hong Kong's distinctive status and advantages. The central government has always handled Hong Kong affairs from a strategic and overall perspective, taking into consideration the fundamental and long-term interests of Hong Kong and the country as a whole. The fundamental interests of Hong Kong are in line with those of the country, and the central government and Hong Kong compatriots share the same aspirations. Hong Kong's close connection with the world market and strong support from the motherland are its distinctive advantages. Such favorable conditions are cherished by the people of Hong Kong and by the central government as well. The central government fully supports Hong Kong in its effort to maintain its distinctive status and edges, to improve its presence as an international financial, shipping, and trading center, to keep its business environment free, open, and regulated, and to maintain the common law, so as to expand and facilitate its exchanges with the world. On the country's journey toward building a modern socialist country in all respects and realizing the rejuvenation of the Chinese nation, the central government believes that Hong Kong will make great contributions.

Fellow Compatriots,

Dear Friends,

Hong Kong compatriots have never been absent in the process, in which the Chinese people and the Chinese nation have realized the great transformation from standing up to growing prosperous and finally to becoming strong. From disarray to good governance, Hong Kong is entering a new phase of becoming more prosperous. The next five years are important for Hong Kong to break new ground and achieve another leap forward. While there are both opportunities and challenges, opportunities outnumber challenges. The central government and people from all sectors of Hong Kong society expect much of the newly inaugurated HKSAR government. People of all ethnic groups across the country wish Hong Kong have promising prospects. For Hong Kong, I have four proposals.

First, Hong Kong should further improve its governance. To promote the development of the HKSAR, it is of urgency to improve Hong Kong's governance system, governance capacity, and governance efficacy. The chief executive and the government of the HKSAR in the driver's seat are the first to be held accountable for the governance of the region. Administrators of Hong Kong should fulfill their commitments, materialize the "one country, two systems" principle with concrete actions, uphold the authority of the Basic Law of the HKSAR and devote themselves to the development of the region. Personnel for public offices should be assessed on both ability and political integrity before they are recruited. Professionals who love both the motherland and Hong Kong with strong governance capabilities and passion for serving the public should be recruited as government staff. Administrators of Hong Kong need to have a new outlook on the motherland and have an international vision in order to make better development plans for the region from an overall and long-term perspective. They need to transform their concepts of governance to balance the relationship between the government and the market so that a capable government serves an efficient market. The HKSAR government needs to strengthen self-governance and improve its conduct to better take on its responsibilities and deliver better performance in ensuring stability and prosperity in Hong Kong.

Second, Hong Kong should continue to create strong impetus for growth. With its special status, Hong Kong enjoys good conditions and broad space for development. The central government fully supports Hong Kong in its effort to seize historic opportunities offered by China's development and actively dovetail itself with the 14th Five-Year Plan (2021-25) and other national strategies such as the development of the Guangdong-Hong Kong-Macao Greater Bay Area and high-quality Belt and Road cooperation. The central government fully supports Hong Kong in carrying out more extensive exchanges and close cooperation with the rest of the world and in attracting entrepreneurs with dreams to realize their ambitions in Hong Kong. The central government also fully supports Hong Kong in taking active yet prudent steps to advance reforms and dismantle the barriers of vested interests in order to unlock enormous creativity and development potential of Hong Kong society.

Third, Hong Kong should earnestly address people's concerns and difficulties in daily life. "Those enjoying benefits and joy of all people should also share their burdens and concerns." As I once said, the people's aspiration for a better life is what we are striving for. Currently, the biggest aspiration of Hong Kong people is to lead a better life, in which they will have more decent housing, more opportunities for starting their own businesses, better education for their children, and better care in their twilight years. We should actively respond to such aspirations. The newly inaugurated HKSAR government should be pragmatic, live up to what the people expect of it, and consider the expectations of the whole society, particularly ordinary citizens, as what it should accomplish foremost. It should be more courageous and adopt more efficient measures to overcome difficulties and forge ahead. It should make sure that all citizens in Hong Kong share more fully and fairly in the fruits of development so that every resident will be convinced that if you work hard, you can improve the life of your own and that of your family.

Fourth, the people of Hong Kong should work together to safeguard harmony and stability. Hong Kong is the home of all its people, and harmony in a family brings success in everything. Through trials and tribulations, now we keenly feel that Hong Kong cannot withstand chaos and will not afford to have any, and we also deeply feel that the development of Hong Kong allows no delay. We must get rid of whatever interference there may be to concentrate our attention on the development of the region. Everyone in Hong Kong, regardless of profession and belief, can be a positive force and do his or her bit for the region's development as long as he or she genuinely supports the principle of "one country, two systems," loves Hong Kong, and abides by the Basic Law and the laws of the special administrative region.

It is my hope that all fellow compatriots in Hong Kong will carry on the mainstream values, which are characterized by the love of both the motherland and Hong Kong as the core and are in conformity with the principle of "one country, two systems," and that they will continue to follow the fine traditions of inclusiveness, seeking common ground while reserving differences, and keeping an unyielding spirit and the courage to strive for success with a view to creating a better future.

We must give special love and care to young people. Hong Kong will prosper only when its young people thrive; Hong Kong will develop only when its young people achieve well-rounded development; and Hong Kong will have a bright future only when its young people have good career prospects. We must guide young people to be keenly aware of the trends in both China and the world and help them cultivate a sense of national pride and enhance their awareness of their status as masters of the country. We must help young people with their difficulties in studies, employment, entrepreneurship, and purchasing of housing, so that more opportunities will be created for their development and accomplishment. We sincerely hope that all of Hong Kong's young people will devote themselves to building Hong Kong into a better home, writing a rewarding chapter of their life with impassioned youth.

Fellow Compatriots,

Dear Friends,

As a Chinese poem goes, "I would like to borrow a pair of wings from the crane to soar up to the sky." China's national rejuvenation has become a historical inevitability, and the successful practice of "one country, two systems" in Hong Kong is an important part of this historic process. We firmly believe that, with the strong backing of the motherland and the solid guarantee provided by "one country, two systems," Hong Kong will surely create a splendid feat on the journey ahead toward the second centenary goal of building China into a modern socialist country in all respects, and will share the glory of the Chinese nation's rejuvenation together with people in the rest of the country.